WILDLY CUTE ANIMALS!

SLOTHS

BY KRISTEN RAJCZAK NELSON

Gareth Stevens
PUBLISHING

Please visit our website, www.garethstevens.com. For a free color catalog of all our high-quality books, call toll free 1-800-542-2595 or fax 1-877-542-2596.

Cataloging-in-Publication Data
Names: Rajczak Nelson, Kristen.
Title: Sloths / Kristen Rajczak Nelson.
Description: Buffalo, NY : Gareth Stevens Publishing, 2026. | Series: Wildly cute animals! | Includes glossary and index.
Identifiers: ISBN 9781482471328 (pbk.) | ISBN 9781482471335 (library bound) | ISBN 9781482471342 (ebook)
Subjects: LCSH: Sloths–Juvenile literature.
Classification: LCC QL737.E2 R35 2026 | DDC 599.3'13–dc23

Published in 2026 by
Gareth Stevens Publishing
2544 Clinton Street
Buffalo, NY 14224

Designer: Jeffrey Taylor
Editor: Kristen Nelson

Photo credits: Series art (tan background) Elena_Design/Shutterstock.com; Cover, p. 1 Milan Zygmunt/Shutterstock.com; p. 5 Harry Collins Photography/Shutterstock.com; p. 7 Lukas Kovarik/Shutterstock.com; p. 9 (top) Milan Zygmunt/Shutterstock.com; p. 9 (top) Swaroop Pixs/Shutterstock.com; p. 11 Diego Grandi/Shutterstock.com; p. 13 Edwin Butter/Shutterstock.com; p. 15 Gualberto Becerra/Shutterstock.com; p. 17 J_K/Shutterstock.com; p. 19 Mark_Kostich/Shutterstock.com; p. 21 Manamana/Shutterstock.com.

Printed in the United States of America

CPSIA compliance information: Batch #CS26GS: For further information contact Gareth Stevens at 1-800-542-2595.

CONTENTS

Boldface words appear in the glossary.

Love the Sloth!

Sloths are wildly cute! In fact, this **mammal** has become so popular in recent years, many kids have sloth T-shirts or stuffies. One kind of these slow-moving animals seems to smile all day long! Let's learn more about sloths.

Where Do They Live?

Sloths live in Central and South America. Their **habitat** is the rainforest, or a forest that gets a lot of rain. It's tropical, or hot and wet, where they live.

Two or Three

The two groups of sloths are the two-toed sloths and the three-toed sloths. They are named for their main difference. Two-toed sloths have two **claws** on their front feet. Three-toed sloths have three. The claws are long and **curved**.

TWO-TOED SLOTH

THREE-TOED SLOTH

Sloths have long **limbs**. They have long fur that comes in shades of brown. Two-toed sloths are bigger than three-toed sloths. They can grow to about 29 inches (74 cm) long and weigh up to 20 pounds (9 kg).

Hanging Around

Sloths spend most of their time in trees. They spend a lot of time hanging upside down! They come down around once a week to go to the bathroom. They may take a swim too!

Low Energy

Sloths seem to sleep a lot because they take a lot of naps. You might not be able to tell when they are awake though. They move slowly. Sloths move so little that **algae** grow on them! Algae help sloths blend in and hide from predators.

Sloths move so slowly because of their **diet**. They eat mostly leaves, buds, stems, and sometimes fruit. This plant matter takes a long time for their body to digest, or break down. It doesn't give sloths much **energy** to move!

Sloths Together

Sloths commonly live alone. Female three-toed sloths may scream at night if they are looking for a **mate**. Two-toed sloth females leave a smell on a tree to draw a mate in. Males may fight over a female. They try to push each other out of the tree!

Sloth mothers have one baby at a time. Babies have claws and all their teeth at birth! A baby sloth crawls to its mother's chest and holds on. It stays there for about the first six months of life.

GLOSSARY

algae: Plantlike living things that are mostly found in water.

claw: An animal's sharp nail.

curved: Not straight, or bent.

diet: The food one usually eats.

energy: Power used to do work.

habitat: The natural place where an animal lives.

limb: An arm or leg.

mammal: A warm-blooded animal that has a backbone and hair, breathes air, and feeds milk to its young.

mate: One of two animals that comes together to make babies.

FOR MORE INFORMATION

BOOKS

Messner, Kate. *Over and Under the Rainforest.* Mankato, MN: Amicus Learning, 2024

Rose, Rachel. *The 10 Cutest Animals.* Minneapolis, MN: Bearport Publishing Company, 2025.

WEBSITES

Brown-Throated Sloths
rainforest-alliance.org/species/sloth/
Discover this wildly cute sloth!

Rainforest Habitat Facts
nationalgeographic.com/nature/habitats/article/rain-forest
Find out more about the rainforest habitat.

Publisher's note to educators and parents: Our editors have carefully reviewed these websites to ensure that they are suitable for students. Many websites change frequently, however, and we cannot guarantee that a site's future contents will continue to meet our high standards of quality and educational value. Be advised that students should be closely supervised whenever they access the internet.

INDEX